Margaret Lee
Moulton

NORMAN ROCKWELL'S
Counting Book

"This is more than just a counting book; it is also a game book because in some of the pictures there are hidden objects and everything counts!"

Norman Rockwell

NORMAN ROCKWELL'S

GLORINA TABORIN

ounting
ook

H·A·R·M·O·N·Y B·O·O·K·S

Book reproduction rights reserved by the artist for all
illustrations with the following exceptions: nos. 2, 18 courtesy
Scouting, U.S.A.; no. 4 courtesy Massachusetts Mutual Life
Insurance Company; no. 14 courtesy Twentieth Century Fox
promotion for "Stagecoach."

Cover detail; nos. 1, 3, 5, 7–11, 13–16, 19; and three April
Fool puzzlers created for Saturday Evening Post covers; nos.
15, 16 prepared for McCall's Magazine; no. 20 courtesy the
Nassau Inn, Princeton.

Library of Congress Catalogue Card Number: 77-78065
ISBN: 0-517-532050
Copyright© 1977 by Harry N. Abrams B.V., The Netherlands

Harmony Books
A division of Crown Publishers, Inc.
One Park Avenue
New York, New York 10016

Printed and bound in Japan

List of Illustrations

1. "Doctor and Doll." Painted for *Saturday Evening Post* cover, March 9, 1929

2. "Diving In." Painted for *Boys' Life (The Boy Scouts' Magazine)* cover, August 1915

3. "Facts of Life." Painted for *Saturday Evening Post* cover, July 14, 1951

4. "Birthday Cake." Advertisement for Massachusetts Mutual Life Insurance Company

5. "Vacation." Painted for *Saturday Evening Post* cover, June 30, 1934

6. "Boy and Dog." Four Seasons calendar, 1958 (autumn)

7. "Cheerleaders." Painted for *Saturday Evening Post* cover, November 25, 1961

8. "St. Nicholas and the Elves." Painted for *Saturday Evening Post* cover, December 2, 1922

9. "Sleeping Fisherman." Painted for *Saturday Evening Post* cover, July 19, 1930

10. "Baby Sitter." Painted for *Saturday Evening Post* cover, November 8, 1947

11. "Discovering Santa." Painted for *Saturday Evening Post* cover, December 29, 1956

12. Promotion for the movie "Stagecoach," 1966

13. "County Agricultural Agent." Illustration painted for *Saturday Evening Post,* July 24, 1948

14. "Newspaper Kiosk." Painted for *Saturday Evening Post* cover, December 20, 1941

15. "Saturday People." Illustration painted for *McCall's Magazine,* October 1966

16. "Stockbridge at Christmas." Illustration painted for *McCall's Magazine,* December 1967

17. "Homecoming G.I." Painted for *Saturday Evening Post* cover, May 26, 1945

18. Boy Scout calendar, 1968

19. "Jester." Painted for *Saturday Evening Post* cover, February 11, 1939

20. "Yankee Doodle." Mural, Nassau Inn, Princeton, N.J.

APRIL FOOL'S DAY PICTURES:

1. Painted for *Saturday Evening Post* cover, April 3, 1943

2. Painted for *Saturday Evening Post* cover, March 31, 1945

3. Painted for *Saturday Evening Post* cover, April 3, 1948

COVER ILLUSTRATION: Detail of *Saturday Evening Post* cover, September 14, 1930

doll

2
feet

3

kittens

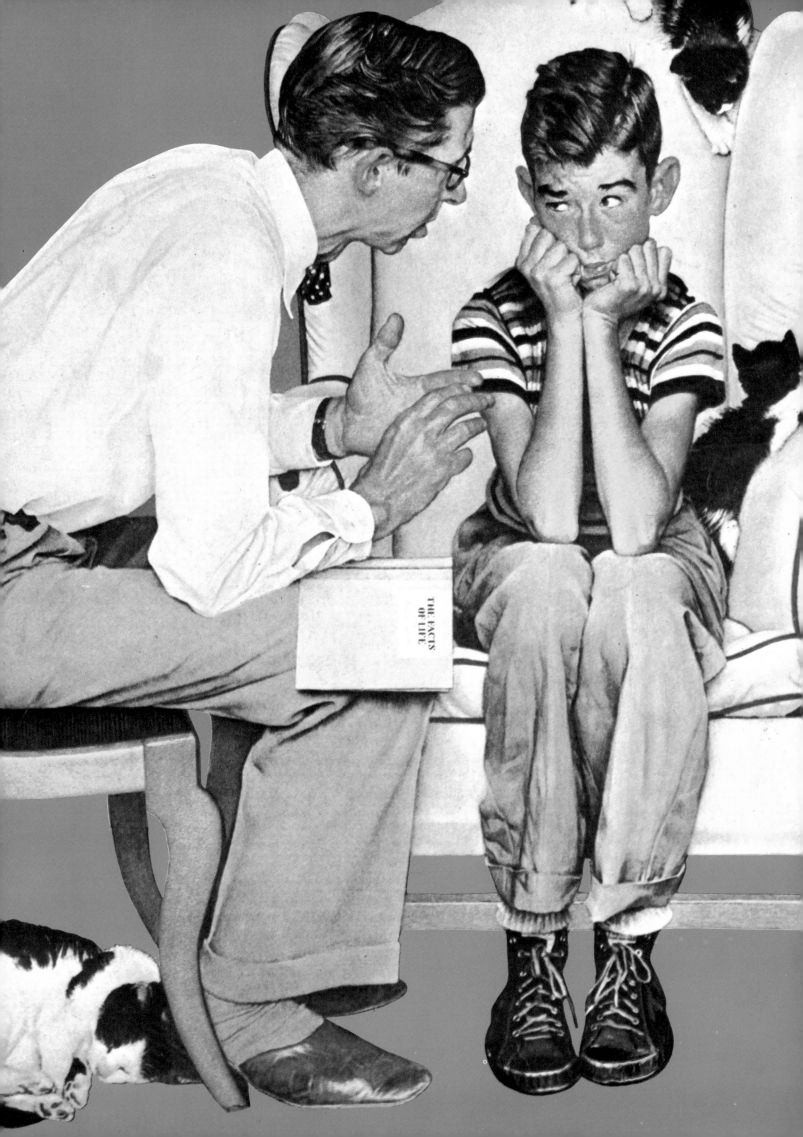

4 birthday hats

5

books

6 dogs

no. 6 ©1958 Brown and Bigelow, St. Paul.

7 cheerleaders

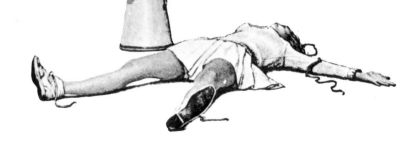

elves

9

fishing poles

10

toes

11

knobs

horses

3 animals

magazines

fingers

automobiles

norman
rockwell

neighbors
greet
returning soldier

boyscouts

19 bells

YANKEE DOODLE CAME TO TOWN · RI

people

HAT · AND CALLED IT MACARONI

NG ON A PONY · STUCK A FEATHER IN HIS

KEEP ON COUNTING!

Now you have counted up to 20. You can count even higher in the next three pictures. They are paintings by Norman Rockwell, and he put many mistakes into them in honor of April Fool's Day. So far, 57 mistakes have been found in the first picture, 43 in the second one, and 51 in the third.

How many mistakes can you count?

Two kinds of molding on cupboard; North American Pileated Woodpecker head on crane's body; Coffeepot spout upside down; Barbed wire instead of clothesline; Insigne on back of fireman's helmet; Green and red lights reversed on ship's lantern; Beast crouched on upper shelf; Cup not hanging by handle; Electric bulbs growing on plant; Head of little girl on man's bust; Rat's tail on chipmunk; Penholder with pencil eraser; Top of brass vase suspended; Face in clock; Candle where kerosene lamp should be; Sampler dated 1216; Winter seen through left window, summer through right; Antique dealer's head on dolls; Nine branches on traditional seven-branch candelabra; Girl's hair in pigtail on one side, loose on other; Titles on books vertical instead of horizontal; Girl's sweater buttoned wrong way; Mouthpiece on both ends of phone; Phone not connected; Goat's head, deer's antlers; No shelf under books; Lace cuff on man's shirt; Five fingers and thumb on girl's hand; Gun barrel in wrong place; Saddle on animal; Potted plant on lighted stove; Girl's purse is a book; Only half a strap on girl's purse; Skunk in girl's arms; Sea gull with crane's legs; Stovepipe missing; Mona Lisa has halo; Mona Lisa facing wrong way; Abraham Lincoln with General Grant's military coat; Stove has April Fool on it; Hoofs instead of feet on doll; Little girl sitting on nothing; Rogers group is combination of soldier from "Our Hero" and girl from "Blushing Bride"; Brass kettle has two spouts; Spur on antique dealer's shoe; Mouse and ground mole conferring; Ground mole's tracks in wooden floor; Dog's head on cat's body; Raccoon's tail on cat's body; Ball fringe standing straight up at angle; Stove minus one leg; Two kinds of floor; Signature reversed; Last name spelled wrong; Flowers growing in floor; Girl's socks don't match; Girl's shoes don't match.

The trout, the fishhook and the water, all on the stairway; Stairway runs behind the fireplace; The mailbox; The faucet; Wallpaper upside down; Wallpaper has two designs; Scissors candlestick; Bacon and egg on the decorative plate; Silhouettes upside down; April-fool clock; The portraits; Ducks in the living room; Zebra looking out of the frame; Mouse looking out of the mantelpiece; A tire for the iron rim of the mantelpiece; Medicine bottle and glass floating in the air; Fork instead of a spoon on the bottle; The old lady's hip pocket; The newspaper in her pocket; Her wedding ring on the wrong hand; Buttons on the wrong side of her sweater; Crown on her head; Stillson wrench for a nutcracker in her hand; Skunk on her lap; She is wearing trousers; She has on ice skates; No checkers on checkerboard; Wrong number of squares on the checkerboard; Too many fingers on old man's hand; Erasers on both ends of his pencil; He is wearing a skirt; He has a bird in his pocket; He is wearing roller skates; He has a hoe for a cane; Billfold on string tied to his finger; Milkweed growing in room; Milk bottle on milkweed; Deer under chair; Dog's paws on deer; Mushrooms; Woodpecker pecking chair; Buckle on man's slipper; Artist's signature in reverse.

Apples on maple tree; Different color apples; Baseball among apples; Pine boughs; Pine cone should point down under bough; Horse-chestnut leaves; Grapes; April 1st comes on Sunday, not Monday; Penguins don't fly; Halo; Nest on phone; Different-color eggs; Phone wire on wrong end of receiver; Different or wrong color butterflies; Books on tree; Castle in landscape; Lighthouse and ship; Earmuffs; Fur collar on velvet jacket; Two different designs on shirt; Shirt buttoned wrong way; Life jacket; Three hands; Cigarette and pipe used at same time; Collar and necktie on bird; Fly-casting reel on bait-casting rod; Cloth patches on waders; Rod upside down; Alligators as roots; Cobra in mandolin; Ribbon on mandolin; Post heading on wrong side of magazine; Snow scene; Horizons different on two scenes; Horns on mouse's head; Animal head on turtle; You're wrong; there are blue lobsters although they are extremely unusual freaks of nature. I once saw one!; Tomato picture on plum can; House slippers on skis; Shells; Dutchman's-breeches; Lady's-slipper; Buttercup; Thimbleweed; Bachelor-buttons; Poison ivy; Signature upside down; Skis without backs; Lead sinkers on line should be below floater; Floater upside down; Red should be at top of floater in right position.